NUREG-1960
Supplement 1

United States Nuclear Regulatory Commission

Protecting People and the Environment

Safety Evaluation Report

Related to the License Renewal of Prairie Island Nuclear Generating Plant Units 1 and 2

Docket Nos. 50-282 and 50-306

Northern States Power Company, a Minnesota Corporation (NSPM)

Office of Nuclear Reactor Regulation

AVAILABILITY OF REFERENCE MATERIALS
IN NRC PUBLICATIONS

United States Nuclear Regulatory Commission

Protecting People and the Environment

NUREG-1960
Supplement 1

Safety Evaluation Report

Related to the License Renewal of Prairie Island Nuclear Generating Plant Units 1 and 2

Docket Nos. 50-282 and 50-306

Northern States Power Company, a Minnesota Corporation (NSPM)

Manuscript Completed: May 2011
Date Published: August 2011

Office of Nuclear Reactor Regulation

ABSTRACT

This document is a supplemental safety evaluation report (SSER) for the license renewal application (LRA) for Prairie Island Nuclear Generating Plant (PINGP), Units 1 and 2, as submitted by Nuclear Management Company (NMC), LLC. By letter dated April 11, 2008, NMC submitted its LRA to the U.S. Nuclear Regulatory Commission (NRC) for renewal of the PINGP operating licenses for an additional 20 years. The NRC staff (the staff) issued a Safety Evaluation Report (SER) Related to the License Renewal of Prairie Island Nuclear Generating Plant Units 1 and 2, dated October 16, 2009 (Agencywide Documents Access and Management System [ADAMS] accession No. ML092890209), which summarizes the results of its review of the LRA for compliance with the requirements of Title 10, Part 54, of the *Code of Federal Regulations* (10 CFR Part 54), "Requirements for Renewal of Operating Licenses for Nuclear Power Plants."

This SSER documents the staff's review of supplemental information provided by the applicant since the issuance of the SER. This information includes annual updates required by 10 CFR 54.21(b), and updated information and commitments in response to recent industry operating experience. The staff informed its evaluation of the applicant's submittals using NUREG-1801, Revision 2, "Generic Aging Lessons Learned (GALL) Report," dated December 2010, which had incorporated this recent industry operating experience.

Paperwork Reduction Act Statement

This NUREG contains information collection requirements that are subject to the Paperwork Reduction Act of 1995 (44 U.S.C. 3501 et seq.). These information collections were approved by the Office of Management and Budget (OMB), approval numbers 3150-0155 and 3150-0011.

Public Protection Notification

The NRC may not conduct or sponsor, and a person is not required to respond to, a request for information or an information collection requirement unless the requesting documents display a currently valid OMB control number.

TABLE OF CONTENTS

ABBREVIATIONS

ACI	American Concrete Institute
ADAMS	Agencywide Documents Access and Management System
AMP	aging management program
AMR	aging management review
ASME	American Society of Mechanical Engineers
ASTM	American Society for Testing of Materials
CAP	corrective action program
CASS	cast austenitic stainless steel
CFR	*Code of Federal Regulations*
CLB	current licensing basis
EQ	environmental qualification or environmentally qualified
GALL	Generic Aging Lessons Learned
GL	generic letter
GSI	Generic Safety Issue
ISI	inservice inspection
kV	kilovolt(s)
LRA	license renewal application
NACE	National Association of Corrosion Engineers
NMC	Nuclear Management Company
NPS	nominal pipe size
NRC	U.S. Nuclear Regulatory Commission
NSPM	Northern States Power Company, a Minnesota Corporation
NUREG	Nuclear Regulation Document

Abbreviations

OE	operating experience
OMB	Office of Management and Budget
PINGP	Prairie Island Nuclear Generating Plant
PWR	pressurized water reactor
PWSCC	primary water stress-corrosion cracking
RAI	request for additional information
RCPB	reactor coolant pressure boundary
SER	safety evaluation report
SRP-LR	Standard Review Plan for Review of License Renewal Applications for Nuclear Power Plants
SSER	supplemental safety evaluation report
SG	steam generator
SSC	system, structure, and component
TLAA	time-limited aging analysis
UFSAR	updated final safety analysis report
V	volt(s)

SECTION 1

INTRODUCTION AND GENERAL DISCUSSION

1.1 Introduction

This document is a supplemental safety evaluation report (SSER) for the license renewal application (LRA) for Prairie Island Nuclear Generating Plant (PINGP), Units 1 and 2, as submitted by Nuclear Management Company (NMC), LLC.

By letter dated September 15, 2008, the U.S. Nuclear Regulatory Commission (NRC) issued an Order approving transfer of operating authority of Facility Operating License Nos. DPR-42 and DPR-60 from NMC, LLC to Northern States Power Company, a Minnesota Corporation (NSPM), for PINGP Units 1 and 2. For the purposes of the SSER, the use of the term "applicant" refers to NMC up to September 15, 2008, and to NSPM on and after September 15, 2008.

By letter dated April 11, 2008, NMC submitted its LRA to the NRC for renewal of the PINGP operating licenses for an additional 20 years. The NRC staff (the staff) issued a Safety Evaluation Report (SER) Related to the License Renewal of Prairie Island Nuclear Generating Plant, Units 1 and 2, dated October 16, 2009 (Agencywide Documents Access and Management System [ADAMS] accession No. ML092890209), which summarizes the results of its review of the LRA for compliance with the requirements of Title 10, Part 54, of the *Code of Federal Regulations* (10 CFR Part 54), "Requirements for Renewal of Operating Licenses for Nuclear Power Plants."

This SSER documents the staff's review of additional information provided by the applicant since the staff's issuance of the SER in October 2009. This information includes annual updates required by 10 CFR 54.21(b), and updated information and commitments in response to recent industry operating experience. The staff informed its evaluation of the applicant's submittals using NUREG-1801, Revision 2, "Generic Aging Lessons Learned (GALL) Report," dated December 2010, which had incorporated this recent industry operating experience. This SSER supplements portions of SER Section 3, Section 4, Appendix A, and Appendix B.

SECTION 2

STRUCTURES, SYSTEMS, AND COMPONENTS

The staff does not have any changes or updates to this section of the safety evaluation report.

SECTION 3

AGING MANAGEMENT REVIEW RESULTS

3.0 Applicant's Use of the Generic Aging Lessons Learned (GALL) Report

3.0.3 Aging Management Programs (AMPs)

3.0.3.1 AMPs Consistent with the GALL Report

3.0.3.1.7 Buried Piping and Tanks Inspection

Summary of Technical Information in the Application. The staff does not have any changes or updates to this section of the safety evaluation report (SER).

Staff Evaluation. The staff does not have any changes or updates to this section of the SER.

Operating Experience. The staff's evaluation of the applicant's proposed Buried Piping and Tanks Inspection Program is documented in 3.0.3.1.7 of the SER issued in October 2009. Given that there have been a number of recent industry events involving leakage from buried or underground piping since the issuance of the SER, the staff identified a need for further information to evaluate the impact that these recent industry events might have on the applicant's Buried Piping and Tanks Inspection Program. The applicant stated that it would provide a supplement to the LRA in its annual update that would address programmatic changes as a result of this recent operating experience (OE).

In its annual update letter dated August 12, 2010, the applicant stated that: (a) a cathodic protection system is maintained with an availability of 90 percent and protects buried piping and tanks; (b) the cathodic protection system rectifier output voltages and currents are checked on a monthly basis; (c) the cathodic protection system is surveyed on an annual basis including component to soil potentials in accordance with NACE (formerly known as National Association of Corrosion Engineers) standards; (d) a risk-informed process will be used to determine specific inspection locations with risk factors including parameters such as coating condition, cathodic protection system assessments, physical layout of the pipe, and consequences of a fluid leak; (e) prior to the period of extended operation, direct inspections will be conducted of 10 feet of pipe in each in-scope system including the cooling water, fire protection, fuel oil, and station and instrument air systems; (f) volumetric examination methods from the inside of the pipe may be substituted where the physical configuration allows for effective assessment; (g) four inspection locations will be inspected in each 10-year period of the license renewal term; and (h) three of the seven in-scope buried tanks will be inspected in each 10-year period commencing 10 years prior to the period of extended operation. The applicant also stated that in October 2009, 27 feet of buried cast iron fire protection piping was excavated for inspection and only minor coating holidays (i.e., minor coating flaws) were noted. The applicant further stated that the locations with holidays were ultrasonically examined and the pipe wall thickness remained above nominal, and all holidays were repaired. The applicant stated that during these excavations, the backfill was noted to consist of sand and small rocks per established engineering specifications.

Based on its review, the staff determined that it did not have sufficient information to find the applicant's annual update letter acceptable. In a conference call conducted on October 4, 2010, the staff requested that the applicant: (a) state whether volumetric examination methods for buried pipe would only be substituted for direct visual examinations when the physical configuration does not allow excavation; (b) state what alternative volumetric examinations beyond ultrasonic examination will be used; (c) if a volumetric examination is used, state what percent of pipe will be inspected; (d) clarify how the completion of buried in-scope tank inspections will be tracked to ensure that all seven tanks will be inspected during the 30 years starting 10 years prior to the period of extended operation; and (e) state the percentage of buried in-scope diesel fuel oil piping that will be inspected during each 10-foot segment inspection.

In its response dated November 5, 2010, the applicant stated that: (a) the program was revised to state that examination from the interior of the pipe would only be used when excavation and direct visual examination is not possible due to plant configuration; (b) only ultrasonic volumetric examinations will be used when inspecting piping from its interior; (c) when using an internal inspection, 25 percent of the buried piping in the affected system will be inspected for loss of material; (d) the program was revised to ensure that the completion of inspections of all seven buried fuel oil tanks are effectively tracked such that all tanks will be inspected during the 30-year period starting 10 years prior to the period of extended operation; and (e) a 10-foot piping segment inspection of buried fuel oil piping would encompass 0.44 percent of all buried in-scope fuel oil piping. The staff noted that inspecting 25 percent of the internal surfaces of the piping with an ultrasonic examination methodology examines a sufficient quantity of piping to establish a reasonable assurance that the piping will meet its current licensing basis (CLB) function(s). The staff also noted that given that the buried fuel oil piping is cathodically protected, coated, and its backfill is acceptable, inspecting 10-foot segments of the piping will establish a reasonable assurance that the piping will meet its CLB function(s).

The license renewal application (LRA) states that the cooling water, station and instrument air, fire protection, and fuel oil systems have buried steel piping. Based on a review of plant-specific OE, the staff noted that the applicant had no documented examples of leaks from these in-scope systems. Inspections of coatings performed during opportunistic inspections of fire protection piping have found only minor holidays in the coatings. Finally, ultrasonic examination of the fire protection piping in the vicinity of the holidays showed that the wall thickness of the piping remained above nominal thickness. The staff finds the applicant's response acceptable because: (a) the rectifiers for the cathodic protection system are monitored on a monthly basis and annual cathodic protection system effectiveness testing is conducted in accordance with standards to assure that cathodic protection availability meets or exceeds 90 percent; (b) all carbon steel piping is coated and current inspections have demonstrated that only minor holidays were present with no degradation of piping below nominal wall thickness; (c) the applicant has appropriate backfill specifications and recent inspections have demonstrated that the backfill quality is consistent with the specifications; (d) at least four excavated visual inspections of at least 10 feet of buried pipe will be conducted in each 10-year period starting 10 years prior to the period of extended operation; (e) each of the seven buried in-scope fuel oil tanks will be inspected during the 30-year period starting 10 years prior to the period of extended operation with three tanks being inspected in each 10-year period; and (f) when internal piping inspections are substituted for excavated direct visual inspections, ultrasonic wall thickness inspections will be used and 25 percent of the buried in-scope portions of the pipe will be inspected. The staff noted that the preventive actions, quantity of inspections, and alternative ultrasonic examinations of the Buried Piping and Tanks Inspection Program are consistent with the corresponding recommendations of GALL Report AMP XI.M41.

Based on its audit and review of the application, and review of the applicant's annual update and supplemental response, the staff finds that OE related to the applicant's program demonstrates that it can adequately manage the detrimental effects of aging on systems, structures, and components (SSCs) within the scope of the program and that implementation of the program has resulted in the applicant taking corrective actions. The staff confirmed that the "operating experience" program element satisfies the criterion in NUREG-1800, Revision 1, "Standard Review Plan for Review of License Renewal Applications for Nuclear Power Plants," (SRP-LR) Section A.1.2.3.10 and, therefore, the staff finds it acceptable.

Updated Final Safety Analysis Report (UFSAR) Supplement. The staff noted that the applicant augmented its UFSAR supplement to state that: (a) a cathodic protection system is provided and maintained in accordance with NACE standards as an additional preventive measure, (b) volumetric examination methods from the interior of components may be substituted for excavation and direct visual examination of the external surfaces of buried components, (c) piping inspection locations are based upon a quantitative risk assessment, (d) a minimum of four inspection locations are inspected every 10-year period of the license renewal term, (e) at least one buried pipe segment in each system will be inspected within the 10 years prior to the period of extended operation, (f) each inspection will include a minimum of 10 feet, and (g) a minimum of three tank inspections are performed once every 10 years with three tanks inspected in the 10-year period preceding the period of extended operation. The staff finds the information added to LRA Section A2.8 acceptable because it aligns the Buried Piping and Tanks Inspection Program with the current staff positions on preventive measures and inspection recommendations for buried pipes and tanks.

Conclusion. The staff does not have any changes or updates to this section of the SER.

3.0.3.1.12 Inaccessible Medium Voltage Cables Not Subject to 10 CFR 50.49 Environmental Qualification Requirements

Summary of Technical Information in the Application. LRA Section B2.1.21 describes the new Inaccessible Medium Voltage Cables Not Subject to 10 CFR 50.49 Environmental Qualification Requirements Program as consistent with GALL AMP XI.E3, "Inaccessible Medium Voltage Cables Not Subject to 10 CFR 50.49 Environmental Qualification Requirements." The applicant stated that this AMP will conduct periodic tests to provide an indication of the condition of the conductor insulation for medium voltage cables within the scope of license renewal exposed to adverse localized environments (i.e., periods of high moisture greater than a few days at a time). The applicant also stated that periodic inspections of the underground medium voltage cable manhole for the accumulation of water (and draining if necessary) will be conducted to minimize prolonged high moisture conditions that promote the growth of water trees.

Staff Evaluation. The staff does not have any changes or updates to this section of the SER.

Operating Experience. The application of GALL AMP XI.E3 to medium voltage cables was based on the OE available at the time Revision 1 of the GALL Report was developed. However, recently-identified industry OE indicates that the presence of water or moisture can be a contributing factor in inaccessible power cable failures at lower operating voltages (400 volts [V] to 2 kilovolts [kV]). Applicable OE was identified in licensee responses to Generic Letter (GL) 2007-01, "Inaccessible or Underground Power Cable Failures that Disable Accident Mitigation Systems or Cause Plant Transients," which included failures of power cables operating at service voltages of less than 2kV where water was considered a contributing factor. The staff has concluded, based on recently-identified industry OE concerning the failure of inaccessible low voltage power cables (400V to 2kV) in the presence of significant moisture, that these cables can potentially experience age-related degradation. The staff noted that the applicant's Inaccessible Medium Voltage Cables Not Subject to 10 CFR 50.49 Environmental Qualification Requirements Program did not address these inaccessible low voltage power cables.

During conference calls with the applicant, the staff requested that the applicant provide the following information:

(1)   Provide a summary of its evaluation of recently-identified industry OE and any plant-specific OE concerning inaccessible low voltage power cable failures within the scope of license renewal (not subject to 10 CFR 50.49 environmental qualification requirements), and how this OE applies to the need for additional aging management activities at its plant for such cables.

(2)   Provide a discussion of how Prairie Island Nuclear Generating Plant (PINGP) will manage the effects of aging on inaccessible low voltage power cables within the scope of license renewal and subject to an aging management review (AMR), with consideration of recently-identified industry OE and any plant-specific OE. The discussion should include an assessment of its AMP description, program elements (i.e., "scope of the program," "parameters monitored or inspected," "detection of aging effects," and "corrective actions"), and the UFSAR summary description to demonstrate reasonable assurance that the intended functions of inaccessible low voltage power cables subject to adverse localized environments will be maintained consistent with the CLB through the period of extended operation.

(3)   Provide an evaluation showing that the Inaccessible Medium Voltage Cables Not Subject to 10 CFR 50.49 Environmental Qualification Requirements Program test and inspection frequencies, including event-driven inspections, incorporate recent industry and plant-specific OE for both inaccessible low and medium voltage cables. Discuss how the Inaccessible Medium Voltage Cables Not Subject to 10 CFR 50.49 Environmental Qualification Requirements Program will ensure that future industry and plant-specific OE will be incorporated into the program such that inspection and test frequencies may be increased based on test and inspection results.

The applicant responded by letter dated November 5, 2010, and stated that in its response to GL 2007-01, "Inaccessible or Underground Power Cable Failures that Disable Accident Mitigation Systems or Cause Plant Transients," dated May 8, 2007, no failures involving low voltage inaccessible cables were identified. The applicant also reviewed more recent PINGP OE to identify any low or medium voltage inaccessible cable failures that may have occurred since the applicant's response to GL 2007-01 (May 2007 through October 2010). The review identified an additional 4kV cable failure not within the scope of license renewal that had experienced water intrusion. The applicant concluded that the failure was due to aging degradation but that water intrusion was not the sole cause of the failure. The cable was replaced. The review of recent OE did not identify any failures of low voltage inaccessible power cables.

Based on plant-specific OE, industry responses to GL 2007-01, and recent U.S. Nuclear Regulatory Commission (NRC) and industry guidance documents, the applicant revised its Inaccessible Medium Voltage Cables Not Subject to 10 CFR 50.49 Environmental Qualification Requirements Program as summarized below:

- The Inaccessible Medium Voltage Cables Not Subject to 10 CFR 50.49 Environmental Qualification Requirements Program is expanded to include 400V to 2kV inaccessible low voltage power cables.

- The exposure to significant voltage (system voltage for more than 25 percent of the time) criterion applied to inaccessible medium voltage cables (2kV to 35kV) is deleted.

- Significant moisture is defined as periodic exposures to moisture that last more than a few days (e.g., cable wetting or submergence in water).

- The Inaccessible Medium Voltage Cables Not Subject to 10 CFR 50.49 Environmental Qualification Requirements Program is expanded to include the inspection of pull boxes with conduit ends containing in-scope inaccessible low and medium voltage power cables for accumulation of water and draining of water, if necessary.

- Manhole and pull box inspection frequencies will be based on actual plant-specific OE with water accumulation, but the inspection frequency will be at least once per year.

- The Inaccessible Medium Voltage Cables Not Subject to 10 CFR 50.49 Environmental Qualification Requirements Program is expanded to include event-driven inspections (e.g., manhole and pull box inspections following a flooding event where river level reaches an elevation where water intrusion might be expected to occur).

- The cable test frequency is revised to at least once every 6 years.

- The initial pull box inspections conducted prior to the period of extended operation will provide baseline information to be used to establish the frequency of future inspections. Additional event-based pull box inspections are to be considered should initial and subsequent inspections indicate water intrusion may be occurring as a result of external events such as heavy rain.

In its discussion of event-driven inspection, the applicant noted that PINGP has one manhole that contains in-scope inaccessible medium voltage cables. The applicant stated that this manhole is constructed approximately 10 feet above the ground water table with a gravel and sand floor designed to drain any water collected. The applicant further stated that the area around the manhole is not subject to significant water accumulation with manhole access located above grade such that significant water intrusion into the manhole is not expected to occur. The applicant also confirmed that manhole inspections performed since September 2007 have shown no signs of water accumulation. Based on the above, the applicant concluded that manhole inspections are not required for a heavy rain event. However, the applicant does include a manhole inspection for river flooding events where river levels reach a sufficient elevation such that water intrusion might be expected. The applicant's program also considers that the periodic inspection frequency may be increased based on inspection results. The staff finds the applicant's assessment acceptable because recent OE shows no history of water accumulation from events such as heavy rain, the manhole design allows accumulated water to drain, and the program considers that the periodic inspection frequency may be increased based on inspection results. In addition, the applicant includes inspections of the in-scope manhole for river flooding events, where river levels reach a sufficient elevation such that water intrusion might be expected.

The applicant also explained how the Inaccessible Medium Voltage Cables Not Subject to 10 CFR 50.49 Environmental Qualification Requirements Program will evaluate future industry and plant-specific OE. The applicant stated that PINGP has a comprehensive OE program that monitors industry issues/events and assesses these for applicability to its own operations. The applicant also stated that it uses the PINGP corrective action program (CAP) to track, trend, and evaluate plant issues/events. The applicant further stated that issues potentially significant to the Inaccessible Medium Voltage Cables Not Subject to 10 CFR 50.49 Environmental Qualification Requirements Program are evaluated by the CAP. If the evaluation shows that changes would enhance program effectiveness, the program is modified as appropriate. The staff finds the applicant's evaluation of industry and plant-specific OE through the CAP to be acceptable because GALL AMP XI.E3 (program elements 7 and 8) concludes that compliance with 10 CFR Part 50, Appendix B is an acceptable means to address corrective actions and is consistent with the guidance of SRP-LR Section A.1.2.3.10, "Operating Experience," and SRP-LR Section A.2, "Quality Assurance," for AMPs (Branch Technical Position IQMB-1).

With the information provided in the applicant's response, the staff finds the Inaccessible Medium Voltage Cables Not Subject to 10 CFR 50.49 Environmental Qualification Requirements Program acceptable with respect to inaccessible low voltage power cables because the applicant has included inaccessible low voltage power cables into this program consistent with industry and plant-specific OE such that there is reasonable assurance that inaccessible low voltage power cables subject to significant moisture will be adequately managed during the period of extended operation. The applicant also revised cable testing frequencies to once every 6 years and manhole and pull box inspections to once every year with added event-driven inspections following elevated river level events for the in-scope manhole and pull boxes. Additional event-driven inspections will be considered for in-scope pull boxes if OE experience indicates that water intrusion is occurring due to external events such as heavy rain. The applicant's incorporation of increased testing and inspection frequencies and event-driven inspections into the Inaccessible Medium Voltage Cables Not Subject to 10 CFR 50.49 Environmental Qualification Requirements Program is consistent with industry, OE and the corresponding recommendations of GALL AMP XI.E3, Revision 2. The elimination of the significant voltage criterion (system voltage for more than 25 percent of the time) is also acceptable because this change expands the scope of the program consistent with industry inaccessible medium voltage cable OE and the corresponding recommendations of GALL AMP XI.E3, Revision 2.

UFSAR Supplement. In LRA Section A2.21, the applicant provided the UFSAR supplement for the Inaccessible Medium Voltage Cables Not Subject to 10 CFR 50.49 Environmental Qualification Requirements Program. By letter dated November 5, 2010, the applicant revised LRA Section A2.21 to include the following: (a) the expansion of the program scope to include inaccessible low voltage (400V to 2kV) power cables, (b) the addition of event-based driven inspections, (c) the revision of cable testing to at least once every 6 years, and (d) the revision of manhole and pull box inspections to at least once every year. The staff reviewed this section and determined that the information in the UFSAR supplement provides an adequate summary description of the program, as required by 10 CFR 54.21(d).

The applicant committed to implement this AMP prior to the period of extended operation and identified it as LRA Commitment No. 17.

Conclusion. The staff does not have any changes or updates to this section of the SER.

3.0.3.1.18 One-Time Inspection

Summary of Technical Information in the Application. The staff does not have any changes or updates to this section of the SER.

Staff Evaluation. The staff's evaluation of the applicant's proposed One-Time Inspection Program is documented in Section 3.0.3.1.18 of the SER issued in October 2009. Based on industry OE, the staff subsequently requested additional information regarding how the selected set of sample components to be inspected would be determined and the subsequent sample size of selected components to be inspected. The staff's evaluation of the additional information submitted, related to the One-Time Inspection Program, is discussed below.

GALL AMP XI.M32, "One-Time Inspection," states in the "detection of aging effects" program element that the inspection includes a representative sample of the system population and, where practical, focuses on the bounding or lead components most susceptible to aging due to time in service, severity of operating conditions, and lowest design margin. The LRA states that the program elements include: (a) determination of the sample size based on an assessment of materials of fabrication, environment, plausible aging effects, and OE; and (b) identification of inspection locations in the system, component, or structure based on the aging effect. However, the LRA did not state how the selected set of sample components would be determined or the size of the sample of components that would be inspected. The staff noted that, due to the uncertainty in determining the most susceptible locations and the potential for aging to occur in other locations, large sample sizes may be required in order to confirm that an aging effect is not occurring. By letter dated November 30, 2010, the staff issued request for additional information (RAI) B2.1.29-1 requesting that the applicant provide specific information regarding how the selected set of components to be sampled will be determined and the size of the sample of components that will be inspected.

In its response dated December 17, 2010, the applicant stated that the components in the One-Time Inspection Program have been placed into four sample groups: (1) components exposed to fuel oil which are being managed for loss of material and cracking, (2) components exposed to treated water or steam which are being managed for loss of material and heat transfer degradation, (3) components exposed to treated water or steam which are being managed for cracking, and (4) components exposed to lubricating or hydraulic oil which are being managed for loss of material and heat transfer degradation. The applicant also stated that for the components exposed to treated water or steam (i.e., groups 2 and 3), a sample size of 20 percent of the population with a maximum of 25 inspections will be established. The applicant further stated that for the components exposed to fuel, lubricating, or hydraulic oil, a reduced sample size will be established because carbon steel, cast iron, copper alloy, and stainless steel materials exposed to fuel, lubricating, or hydraulic oil should not experience loss of material except in locations where water or other contaminants are present for an extended period of time.

Regarding the components exposed to fuel oil, the applicant stated that plant-specific OE has not identified any problems with water, particulates, or biological fouling in its fuel oil and that the fuel oil storage tanks represent the low points in the system. The applicant also stated that it will inspect four of its six fuel oil tanks, four of its seven fuel oil day tanks, one of its two clean fuel oil leakage collection tanks, and six other locations which include a sample of each material and environment combination within the system. The applicant further stated that these inspections are adequate to ensure that the low points and stagnant areas are inspected while minimizing inspections of less susceptible components.

Regarding the components exposed to lubricating or hydraulic oil, the applicant stated that the 1,048 components in the group are associated with 11 systems in which the collection areas for contaminants or water pooling are readily identifiable. The applicant also stated that it will inspect 17 of the 1,048 components in the group such that a sample from each material and environment combination is inspected and is adequate to ensure low points and stagnant areas are checked.

The staff reviewed the applicant's response and noted that fuel, lubricating, or hydraulic oil do not create an environment conducive to loss of material unless water or other contaminants are able to collect in the low or stagnant points in the systems and, therefore, the most susceptible locations are often readily identifiable in these systems. The staff finds the applicant's response acceptable because: (a) the applicant's selected set of components to be sampled will be based on material and environment combinations; (b) the sample locations for components exposed to treated water or steam will focus on the leading indicator components and include an appropriate sample size; (c) loss of material is not expected for components exposed to fuel, lubricating, or hydraulic oil except in areas where water or other contaminants are able to collect; and (d) the applicant has chosen inspection locations for these groups of components which focus on these areas. The staff noted that the applicant's sampling methodology is consistent with the corresponding recommendations in GALL Report AMP XI.M32, Revision 2. The staff's concern described in RAI B2.1.29-1 is resolved.

Operating Experience. The staff does not have any changes or updates to this section of the SER.

UFSAR Supplement. The staff does not have any changes or updates to this section of the SER.

Conclusion. The staff does not have any changes or updates to this section of the SER.

3.0.3.1.19 One-Time Inspection of ASME Code Class 1 Small-Bore Piping

Summary of Technical Information in the Application. The staff does not have any changes or updates to this section of the SER.

Staff Evaluation. The staff's evaluation of the applicant's proposed One-Time Inspection of ASME Code Class 1 Small-Bore Piping Program is documented in Section 3.0.3.1.19 of the SER issued in October 2009. Based on recently-identified industry OE associated with cracking in American Society of Mechanical Engineers (ASME) Class 1 small-bore socket welds, the staff subsequently requested additional information regarding the inspections that would be performed by the applicant's One-Time Inspection of ASME Code Class 1 Small-Bore Piping Program. The staff's evaluation of the additional information submitted is discussed below.

By letters dated August 12 and November 5, 2010, the applicant submitted an annual update of its LRA and supplemental information to its LRA, respectively. The staff reviewed the information regarding the applicant's One-Time Inspection of ASME Code Class 1 Small-Bore Piping Program and noted that the program only addresses Class 1 small-bore full penetration welds but did not adequately address volumetric examination of socket welds. Specifically, nominal pipe size (NPS) 1-inch Class 1 small-bore socket welds were excluded from the applicant's inspection population. By letter dated November 30, 2010, the staff issued RAI B2.1.30 requesting that the applicant supplement its program to incorporate NPS 1-inch socket welds in its inspection program and provide adequate inspection sampling.

In its response dated December 17, 2010, the applicant provided supplemental information to its One-Time Inspection of ASME Code Class 1 Small-Bore Piping Program. The applicant stated that the program scope is revised to include NPS 1-inch Code Class 1 piping. The staff noted that since the scope of the applicant's inspection program was amended to include NPS 1-inch Code Class 1 piping, which is consistent with the "scope of the program" program element of GALL AMP XI.M35, Revision 2, "One-Time Inspection of ASME Code Class 1 Small-Bore Piping," the staff finds it acceptable.

In the supplements dated November 5 and December 17, 2010, the applicant discussed plant-specific OE regarding Code Class 1 small-bore piping and stated that it has not experienced any failures. The applicant stated that it would perform volumetric examinations of "3% of the Code Class 1 small-bore socket welds, up to a maximum of ten welds," at each of the applicant's units. The staff noted that each of the applicant's units has more than 30 years of operation and there have been no Class 1 small-bore piping failures in its plant-specific OE. Based on recommendations of GALL AMP XI.M35, Revision 2, if an applicant has never experienced a failure in its Class 1 small-bore piping and has more than 30 years of operation, the inspection sampling should be at least 3 percent of the weld population or a maximum of 10 welds of each weld type for each operating unit. The staff finds that the applicant's proposed inspection sample size as described in its letter dated November 5, 2010, is consistent with the "detection of aging effects" program element of GALL AMP XI.M35, Revision 2, and is, therefore, acceptable.

The applicant also stated that if an acceptable volumetric technique is not available for the site to perform these inspections, it would perform a destructive examination. The applicant further stated that, "Each destructive weld examination will be considered equivalent to performing two volumetric weld examinations." Based on the recommendations of GALL AMP XI.M35, Revision 2, an applicant may take credit for each socket weld that is destructively examined as being equivalent to volumetrically examining two socket welds because more information can be obtained from a destructive examination than from a nondestructive examination. The staff finds the applicant's proposed alternative to perform destructive examinations, in lieu of volumetric examinations, acceptable because it is consistent with the recommendations of the "detection of aging effects" program element of GALL AMP XI.M35, Revision 2, and more information can be obtained from a destructive examination than from nondestructive examination.

Regarding the implementation schedule for the One-Time Inspection of ASME Code Class 1 Small-Bore Piping Program, the applicant stated that it would perform volumetric examination on five socket welds (two at Unit 1, three at Unit 2) or destructive examination of two socket welds at each unit prior to entering the period of extended operation. The applicant further stated that:

> Because of the limited number of refueling outages remaining prior to the period of extended operation, and in order to allow orderly planning and scheduling of plant resources and outage workload, the additional socket weld examinations, required as the result of applying the 3% (up to a maximum of 10 welds)

sampling criteria to the increased weld population, will be performed within three years of entering the period of extended operation.

Since the applicant will be entering the period of extended operation on August 9, 2013, for Unit 1 and October 29, 2014, for Unit 2, the staff noted that the first inspection will be performed within 3 years prior to entering the period of extended operation and all inspections will be completed within 3 years after entering the period of extended operation. The staff finds the applicant's proposal consistent with the recommendations of the "detection of aging effects" program element of GALL AMP XI.M35, Revision 2, regarding timely implementation of the small-bore piping inspections and is, therefore, acceptable.

The staff noted that the applicant uses its risk-informed methodology for sample selection to ensure the most susceptible and risk-significant welds are selected. The "detection of aging effects" program element of GALL AMP XI.M35, Revision 2, recommends a methodology that selects the most susceptible and risk-significant welds to inspect. The staff finds that the sample selection methodology is consistent with the "detection of aging effects" program element of GALL AMP XI.M35, Revision 2, and is, therefore, acceptable.

The staff determined that aging management of Code Class 1 small-bore piping is adequately addressed because the scope of the program, number of welds to be inspected, the selection methodology, and the timely implementation of the small-bore piping inspection is consistent with the recommendations in the GALL Report.

Based on its review, the staff finds the One-Time Inspection of ASME Code Class 1 Small-Bore Piping Program consistent with the program elements of GALL AMP XI.M35 and, therefore, acceptable.

Operating Experience. The staff does not have any changes or updates to this section of the SER.

UFSAR Supplement. The staff does not have any changes or updates to this section of the SER.

Conclusion. The staff does not have any changes or updates to this section of the SER.

3.0.3.2 AMPs Consistent with the GALL Report with Exceptions or Enhancements

3.0.3.2.15 Selective Leaching of Materials Program

Summary of Technical Information in the Application. The staff does not have any changes or updates to this section of the SER.

Staff Evaluation. The staff's evaluation of the applicant's proposed Selective Leaching of Materials Program is documented in Section 3.0.3.2.15 of the SER issued in October 2009. Based on industry OE, the staff subsequently requested additional information regarding how the selected set of sample components to be inspected would be determined and the subsequent sample size of selected components to be inspected. The staff's evaluation of the additional information submitted related to the Selective Leaching of Materials Program is discussed below.

GALL AMP XI.M33, "Selective Leaching of Materials," states in the "scope of the program" program element that the program includes a one-time visual inspection and hardness measurement of a selected set of sample components to determine whether loss of material due to selective leaching does not occur for the period of extended operation. However, the LRA did not state how the selected set of sample components would be determined or the size of the sample of components that would be inspected. The staff noted that due to the uncertainty in determining the most susceptible locations and the potential for aging to occur in other locations, large sample sizes may be required in order to confirm that selective leaching is not occurring. By letter dated November 30, 2010, the staff issued RAI B2.1.36-2 requesting that the applicant provide specific information regarding how the selected set of components to be sampled will be determined and the size of the sample of components that will be inspected.

In its response dated December 17, 2010, the applicant stated that the sample groups will be based on the materials of fabrication (e.g., gray cast iron and copper alloy with greater than 15 percent zinc), and that a sample size of 20 percent of the population, with a maximum of 25 inspections, will be established for each sample group. The applicant also stated that: (a) the sample locations will be developed to ensure that a representative sample of material and environment combinations is selected such that at least one component from each susceptible material and environment combination is inspected and (b) it will focus on inspecting the "leading indicator" or most susceptible components. The staff finds the applicant's response acceptable because the applicant's sampling methodology: (a) ensures a representative sample of material and environment combinations is considered, (b) ensures sample locations will focus on the leading indicator components, and (c) includes an appropriate sample size. The staff noted that the applicant's sampling methodology is consistent with recommendations in GALL Report AMP XI.M33, Revision 2. The staff's concern described in RAI B2.1.36-2 is resolved.

Exception 1. The staff does not have any changes or updates to this section of the SER.

Operating Experience. The staff does not have any changes or updates to this section of the SER.

UFSAR Supplement. The staff does not have any changes or updates to this section of the SER.

Conclusion. The staff does not have any changes or updates to this section of the SER.

3.0.3.2.17 Structures Monitoring

Summary of Technical Information in the Application. The staff does not have any changes or updates to this section of the SER.

Staff Evaluation. The staff's evaluation of the applicant's proposed Structures Monitoring Program is documented in Section 3.0.3.2.17 of the SER issued in October 2009. Subsequently, the staff requested additional information regarding the applicant's acceptance criteria. The staff's evaluation of the additional information submitted related to the Structures Monitoring Program is discussed below.

GALL AMP XI.S6, "Structures Monitoring Program," states that American Concrete Institute (ACI) 349.3R is an acceptable basis for selection of parameters monitored, detection of aging effects, and acceptance criteria. The LRA states that the applicant's program incorporates inspection guidance based on recommendations contained in ACI 349.3R; however, it does not clearly state that the acceptance criteria align with those in ACI 349.3R. By letter dated November 30, 2010, the staff issued RAI B2.1.38 requesting the applicant to confirm that quantitative acceptance criteria consistent with those in ACI 349.3R are included in the applicant's Structures Monitoring Program, or justify any changes.

By letter dated December 17, 2010, the applicant stated that its Structures Monitoring Program includes quantitative acceptance criteria which are consistent with those in Chapter 5 of ACI 349.3R. The staff reviewed the applicant's response and found it acceptable because the applicant clarified that its acceptance criteria was in alignment with the quantitative criteria recommended in ACI 349.3R. The staff also noted that the applicant's acceptance criteria is consistent with the corresponding recommendations in GALL Report AMP XI.S6, Revision 2. The staff's concern described in RAI B2.1.38 is resolved.

Operating Experience. The staff does not have any changes or updates to this section of the SER.

UFSAR Supplement. The staff does not have any changes or updates to this section of the SER.

Conclusion. The staff does not have any changes or updates to this section of the SER.

3.1 Aging Management of Reactor Vessel, Internals, and Reactor Coolant System

3.1.2 Staff Evaluation

3.1.2.1 AMR Results Consistent with the GALL Report

3.1.2.1.6 Cracking Due to Primary Water Stress-Corrosion Cracking (Steam Generator Divider Plate)

LRA Table 3.1.1, item 81, addresses cracking due to primary water stress-corrosion cracking (PWSCC) for the nickel-alloy or nickel-alloy clad steam generator (SG) divider plate exposed to reactor coolant. LRA Table 3.1.1, item 82, states that the SG primary side divider plates for both Units 1 and 2 are fabricated from nickel-alloy. The applicant credited the Water Chemistry Program to manage cracking due to PWSCC in nickel-alloy SG divider plates exposed to reactor coolant, consistent with the GALL Report.

The staff noted that, from foreign OE in SGs with a similar design to that of the applicant's SGs, extensive cracking due to PWSCC has been identified in SG divider plate assemblies fabricated from Alloy 600, even with proper primary water chemistry. The staff noted that, specifically, cracks have been detected in the stub runner, very close to the tubesheet/stub runner weld and with depths of almost a quarter of the divider plate thickness. Therefore, the staff noted that the Water Chemistry Program may not be effective in managing the aging effect of cracking due to PWSCC in SG divider plate assembly components fabricated from Alloy 600 and its associated weld metals.

The staff noted that these SG divider plate assembly cracks could affect adjacent items that are part of the reactor coolant pressure boundary (RCPB), such as the tubesheet and the channel head, if they propagate to the boundary with these items. The staff further noted that PWSCC cracks in the divider plate assemblies fabricated from Alloy 600 and its associated weld metals could propagate to the tubesheet cladding with possible consequences to the integrity of the tube-to-tubesheet welds. Furthermore, for the channel head, the PWSCC cracks in the divider plate assemblies could propagate to the SG triple point and potentially affect the pressure boundary of the SG channel head.

The staff reviewed the applicant's UFSAR and noted that the Unit 1 UFSAR Table 4.1-1 states that the divider plate is made with Alloy 690 for the Unit 1 replacement Framatome SGs. For Unit 2, UFSAR Table 4.1-1 describes the construction materials for the original Westinghouse SGs. However, the staff noted that there is no information about the construction materials for the divider plate assembly for the Unit 2 SGs.

The staff held conference calls on October 4 and October 27, 2010, with the applicant to discuss and clarify the staff's concern. The staff asked the applicant: (1) to clarify whether all the components for Unit 1 replacement SG divider plate assemblies, including the welds within these assemblies as well as to the channel head and to the tubesheet, are fabricated from Alloy 690 or its associated weld materials and to describe the construction materials of Unit 2 SG divider plate assemblies, including the welds within these assemblies as well as to the channel head and to the tubesheet; and (2) if any constitutive/weld material of the SG divider plate assemblies is susceptible to cracking (e.g., Alloy 600 or its associated weld materials), to describe an inspection program (examination technique and frequency) to ensure that there are no cracks propagating into other items which are part of the RCPB (e.g., tubesheet and channel head) that could challenge the integrity of those adjacent items.

By letter dated November 5, 2010, the applicant amended its LRA to provide supplemental information related to SG divider plate materials and inspections. The applicant described the materials used in the fabrication of the Unit 1 replacement SG divider plates and associated welds, which are: Inconel Alloy 690 for the divider plate, Inconel weld material 152 for the stub weld buildup on the tubesheet, Inconel weld material 152 for the weld material between the stub runner and the divider plate, and 308L-316 stainless steel for the weld material between the channel head and the divider plate. The applicant also stated that the Unit 2 replacement SGs are being fabricated from the same materials.

The applicant stated that because the Unit 1 and Unit 2 replacement SG divider plates and associated welds use Alloy 690 and its associated weld material, or austenitic stainless steel, no inspection of the replacement SG divider plates and associated welds is required. The staff considers that the use Alloy 690 and its associated weld material prevent cracking due to PWSCC for the SG divider plate assembly components and associated weld metals because it contains a higher chromium content.

The applicant further stated that the materials used in the fabrication of the divider plates and associated welds in the Unit 2 original SGs are Inconel Alloy 600 for the divider plate, and Inconel 182 for the weld material (divider plate to tubesheet and channel head). The applicant also clarified that the replacement SGs for Unit 2 are currently in fabrication and are scheduled for installation during the last Unit 2 refueling outage prior to the period of extended operation, and no inspection of the original Unit 2 SG divider plates and associated welds is deemed necessary prior to the period of extended operation. However, in case the original Unit 2 SGs are not replaced prior to entry into the period of extended operation, the applicant committed (Commitment No. 45) to the following:

If the original PINGP Unit 2 steam generators are not replaced prior to entry into the period of extended operation, NSPM will perform an inspection of each PINGP Unit 2 steam generator, prior to the period of extended operation, to assess the condition of the divider plates and associated welds. The examination technique(s) will be capable of detecting PWSCC in the divider plates and associated welds.

Based on its review, the staff finds the applicant's proposal and associated Commitment No. 45 acceptable because the applicant's replacement SGs will not include any Alloy 600 or its associated weld materials in the divider plate assemblies when the SGs enter the period of extended operation. Moreover, in case the applicant is not able to install replacement SGs before Unit 2 enters into the period of extended operation, the staff noted that the applicant would assess, prior to the period of extended operation, the condition of the divider plate assemblies, which contain Alloy 600 and its associated weld materials, in each Unit 2 original SG by inspection with appropriate examination techniques.

Based on the programs and commitments identified, the staff concludes that the applicant's methodology for aging management meets SRP Section 3.1.2.2.11.1, Revision 2 criteria. The staff concludes that the applicant has demonstrated that the effects of aging for these components will be adequately managed so that their intended function(s) will be maintained consistent with the CLB during the period of extended operation, as required by 10 CFR 54.21(a)(3).

3.1.2.2 AMR Results Consistent with the GALL Report for Which Further Evaluation is Recommended

3.1.2.2.16.1 Cracking due to Stress-Corrosion Cracking and Primary Water Stress-Corrosion Cracking

SRP-LR Section 3.1.2.2.16.1 is associated with AMR items 34 and 35 in Table 1 of the GALL Report, Volume 1, and LRA Table 3.1.1, items 3.1.1-34 and 3.1.1-35. The staff's evaluation of LRA Table 3.1.1, item 3.1.1-34, is documented in Section 3.1.2.2.16.1 of the SER issued in October 2009. Subsequently, the staff noted that AMR item 35 in Table 1, of the GALL Report, Volume 1, may be applicable to recirculating SGs. Therefore, the staff further considered LRA Table 3.1.1, item 3.1.1-35, which states that it is not applicable because this line applies only to once-through SGs and not to recirculating SGs, which are used at the applicant's plant.

SRP-LR Section 3.1.2.2.16.1 identifies that cracking due to PWSCC could occur on the primary coolant side of pressurized water reactor (PWR) steel SG tube-to-tubesheet welds made of cladding with nickel alloy. The GALL Report recommends ASME Section XI Inservice Inspection (ISI) and control of water chemistry to manage cracking due to PWSCC and recommends no further AMR for PWSCC of nickel alloy if the applicant complies with applicable NRC Orders and provides a commitment in the UFSAR supplement to implement applicable: (1) bulletins and GLs, and (2) staff-accepted industry guidelines. In GALL Report Revision 1, Volume 2, cracking due to PWSCC is addressed in item IV.D2-4 and is applicable only to once-through SGs, but not to recirculating SGs.

The staff noted that ASME Code Section XI does not require any inspection of the tube-to-tubesheet welds. In addition, no specific NRC Orders or bulletins address inspection requirements for these welds. The staff's concern is that, if the tubesheet cladding is Alloy 600, or the associated weld material is Alloy 600, the region of the autogenous tube-to-tubesheet weld may have insufficient chromium content to prevent initiation of PWSCC, even when the SG tubes are made from Alloy 690TT. Consequently, a crack initiated in this region, close to a tube, may propagate into or through the weld, causing a failure of the weld and of the RCPB. This could occur in recirculating SGs such as those used at both of the applicant's units. Therefore, unless the NRC has approved a redefinition of the RCPB in which the autogenous tube-to-tubesheet weld is no longer included, or the tubesheet cladding and welds are not susceptible to PWSCC, the staff considers that the effectiveness of the primary water chemistry program should be verified to ensure PWSCC cracking is not occurring.

UFSAR Table 4.1-1 states that the Unit 1 replacement Framatome Model 56/19 SG tubes are fabricated from Alloy 690TT and the cladding for the tubesheets from Alloys 82 and 182. Furthermore, the Unit 2 Westinghouse Model 51 SG tubes are fabricated from Alloy 600MA and the cladding for the tubesheets is Inconel. UFSAR Section 4.3.2.4 further states that the NRC has approved an amendment to its technical specifications which allows Unit 2 SG tubes to remain in service if the required length of hard roll expansion is intact above the highest degradation in the tubesheet crevice region (F* Alternate Repair Criteria). However, the staff noted that the applicant will replace Unit 2 SGs before the period of extended operation. Therefore, the staff does not have sufficient information about the configuration of the replacement SGs tube-to-tubesheet welds and the necessity to manage the potential aging effect of cracking due to PWSCC in these welds.

By letter dated November 30, 2010, the staff issued RAI 3.1.2.2.16 requesting that, for the Unit 1 SGs, the applicant provide either a plant-specific AMP that will complement the primary water chemistry program, in order to verify the effectiveness of the primary water chemistry program and ensure that cracking due to PWSCC is not occurring in tube-to-tubesheet welds, or a rationale for why such a program is not needed. For the Unit 2 SGs, the staff requested that the applicant clarify: (1) when it will replace the original SGs, especially whether this replacement will occur before or after the period of extended operation; (2) describe whether the tubesheet cladding and the tube-to-tubesheet welds of the future replacement SGs are susceptible to PWSCC; and (3) provide the materials of construction. If these materials are potentially susceptible to PWSCC (e.g., Alloy 600 and/or its associated weld metals), the staff requested that the applicant provide an AMP that will verify the effectiveness of the primary water chemistry program and will ensure that cracking due to PWSCC is not occurring in the tube-to-tubesheet welds.

In its response dated December 17, 2010, the applicant stated that it would perform a one-time inspection of a representative number of tube-to-tubesheet welds in each Unit 1 SG to determine if PWSCC is present, and if weld cracking is identified, it would implement corrective actions, including an evaluation of the degradation and the implementation of routine inspections of the tube-to-tubesheet welds for the remaining life of the Unit 1 and Unit 2 replacement SGs. The applicant further stated that the replacement Unit 1 SGs have accumulated approximately 6 years of service time since having been replaced in 2004. Considering this limited service time, the applicant stated that the SG tube-to-tubesheet weld inspections would be performed during the first Unit 1 refueling outage after the SGs have reached 20 years of service. For Unit 2, the applicant stated that, as discussed in its letter L-PI-10-109 dated November 5, 2010, the replacement Unit 2 SGs are currently in fabrication and are scheduled for installation during the last Unit 2 refueling outage prior to the period of extended operation. The applicant stated that the Unit 2 replacement SG tubesheets are clad with Inconel Alloy 600 material, and as such, the tube-to-tubesheet welds are susceptible to PWSCC. The applicant clarified that, based on the current schedule for their installation, the Unit 2 replacement SGs will have only accumulated approximately 21 years of service time at the end of the period of extended operation. Therefore, because of this limited service time, the applicant stated that no one-time inspection of the Unit 2 tube-to-tubesheet welds is deemed necessary during the period of extended operation. However, the applicant stated that, as committed to in its response to RAI 3.1.2.2.16 for Unit 1 (Commitment No. 46 below), in the event that the one-time inspection of the Unit 1 SG tube-to-tubesheet welds identifies cracking, an AMP would be established to perform routine tube-to-tubesheet weld inspections for the remaining life of the Unit 1 and Unit 2 replacement SGs.

In response to the staff's concern, the applicant committed (Commitment No. 46) to the following:

> A one-time inspection of a representative number of tube-to-tubesheet welds in each Unit 1 steam generator will be performed to determine if primary water stress corrosion cracking (PWSCC) is present. The tube-to-tubesheet weld inspections will be performed during the first Unit 1 refueling outage after the Unit 1 steam generators have reached 20 years of service. If weld cracking is identified:
>
> a. The condition will be resolved through repair or engineering evaluation to justify continued service, as appropriate, and
>
> b. An aging management program will be established to perform routine tube-to-tubesheet weld inspections for the remaining life of the Unit 1 and Unit 2 replacement steam generators.

Based on its review, the staff finds the applicant's response to RAI 3.1.2.2.16 and associated Commitment No. 46 acceptable because the applicant will manage the aging effect of cracking due to PWSCC in the SG tube-to-tubesheet welds by initially implementing a one-time inspection on a representative number of tube-to-tubesheet welds of each Unit 1 SG to determine if PWSCC is present. The staff finds the timing of this one-time inspection for the Unit 1 SGs acceptable because, at the time of the inspection, the SGs will have been in operation for between 20 and 25 years, and it is unlikely that significant detrimental PWSCC will have initiated before this time period. For Unit 2, the staff finds the applicant's rationale for not performing an inspection of the SG tube-to-tubesheet welds during the period of extended operation acceptable because of the limited operating time of these SGs at the end of the period of extended operation. Moreover, if the applicant is not able to install the replacement SGs before Unit 2 enters into its period of extended operation, the staff noted that the NRC has approved an amendment to its technical specifications (F* Alternate Repair Criteria) for the original SGs; therefore, aging management for PWSCC of the original SG tube-to-tubesheet welds is not necessary since these welds are no longer included in the RCPB. The staff also noted that, if the aging effect is identified by the Unit 1 inspection, the applicant would take corrective actions, including an evaluation of the degradation and the implementation of routine inspections of the tube-to-tubesheet welds for the remaining life of both units' replacement SGs. The staff's concern described in RAI 3.1.2.2.16 is resolved.

Based on the programs identified, the staff concludes that the applicant's programs meet SRP-LR, Revision 1, Section 3.1.2.2.16.1 and SRP Section 3.1.2.2.11.2, Revision 2 criteria. For those line items that apply to LRA Section 3.1.2.2.16.1, the staff determines that the LRA is consistent with the GALL Report, Revision 2 and that the applicant has demonstrated that the effects of aging will be adequately managed so that the intended function(s) will be maintained consistent with the CLB during the period of extended operation, as required by 10 CFR 54.21(a)(3).

SECTION 4

TIME-LIMITED AGING ANALYSES

4.3 Metal Fatigue

4.3.3 Environmentally-Assisted Fatigue (Generic Safety Issue [GSI]-190)

4.3.3.1 Summary of Technical Information in the Application

The staff does not have any changes or updates to this section of the SER.

4.3.3.2 Staff Evaluation

The staff's review of the applicant's evaluation of environmentally-assisted fatigue is documented in Section 4.3.3.2 of the SER issued in October 2009. Subsequently, the staff noted that the applicant's plant-specific configuration may contain locations that should be analyzed for the effects of the reactor coolant environment other than those generic locations identified in NUREG/CR-6260. The staff's evaluation of the additional information submitted related to environmentally-assisted fatigue is discussed below.

By letter dated November 30, 2010, the staff issued RAI 4.3.3 requesting the applicant to confirm and justify that the locations selected for environmentally-assisted fatigue analyses, consistent with NUREG/CR-6260, are the most limiting and bounding for the plant. Furthermore, if these locations are not the most limiting and bounding for the plant, the applicant was requested to clarify the locations that require an environmentally-assisted fatigue analysis and the actions that will be taken for these additional locations. If the most limiting location consists of nickel alloy, the applicant was asked to clarify whether the methodology that it would use for environmentally-assisted fatigue is consistent with the NUREG/CR-6909 methodology for nickel alloy.

In its response dated December 17, 2010, the applicant discussed the bases for the selection of plant-specific component locations as the limiting locations that were evaluated for environmentally-assisted fatigue. The applicant also committed (Commitment No. 47) to the following:

> NSPM will perform a review of the design basis ASME Class 1 fatigue evaluations to determine whether the NUREG/CR-6260 components that have previously been evaluated for the effects of reactor coolant environment on fatigue life are the limiting components for the PINGP design.
>
> a. If a more limiting component(s) is identified, the most limiting component will be evaluated for the effects of the reactor coolant environment on fatigue usage.
>
> b. If the limiting component identified consists of nickel alloy, the methodology used to perform the environmentally assisted fatigue calculation for nickel alloy will be consistent with NUREG/CR-6909, or otherwise justified.

Based on its review, the staff finds the applicant's response to RAI 4.3.3 and Commitment No. 47 acceptable because: (a) the applicant will review its design basis ASME Code Class 1 fatigue evaluations to determine whether the NUREG/CR-6260 components are the limiting components for the applicant's design; (b) if more limiting component(s) are identified, the applicant will perform environmentally-assisted fatigue analyses for the most limiting component; (c) a methodology consistent with NUREG/CR-6909 will conservatively be used in the evaluation if the limiting component identified consists of nickel alloy; and (d) Commitment No. 47 is consistent with the recommendations in SRP-LR, Revision 2, Sections 4.3.2.1.3 and 4.3.3.1.3, and GALL Report AMP X.M1, Revision 2, "Fatigue Monitoring," to consider environmental effects for the NUREG/CR-6260 locations, at a minimum.

4.3.3.3 UFSAR Supplement

The staff does not have any changes or updates to this section of the SER.

4.3.3.4 Conclusion

The staff does not have any changes or updates to this section of the SER.

SECTION 5

REVIEW BY THE ADVISORY COMMITTEE ON REACTOR SAFEGUARDS

The staff has provided the Advisory Committee on Reactor Safeguards with a copy of this supplemental safety evaluation report.

SECTION 6

CONCLUSION

The staff concludes that the additional information provided by Northern States Power Company, a Minnesota Corporation (NSPM), does not alter the conclusion proffered in the safety evaluation report issued in October 2009 and that the requirements of Title 10, Section 54.29(a) of the *Code of Federal Regulations* (10 CFR 54.29(a)) have been met.

APPENDIX A

PRAIRIE ISLAND NUCLEAR GENERATING PLANT LICENSE RENEWAL COMMITMENTS

During the review of the Prairie Island Nuclear Generating Plant (PINGP), Units 1 and 2, license renewal application (LRA) by the staff of the U.S. Nuclear Regulatory Commission (NRC) (the staff), Northern States Power Company, a Minnesota Corporation (NSPM) (the applicant) made commitments related to aging management programs (AMPs) to manage aging effects for structures and components.

The following table contains the final complete list of these commitments along with the implementation schedules and sources for each commitment.

Appendix A

APPENDIX A: PINGP LICENSE RENEWAL COMMITMENTS

Commitment Number	Commitment	UFSAR Supplement Section/LRA Section	Enhancement or Implementation Schedule
1	Each year, following the submittal of the PINGP License Renewal Application and at least three months before the scheduled completion of the NRC review, NSPM will submit amendments to the PINGP application pursuant to 10 CFR 54.21(b). These revisions will identify any changes to the Current Licensing Basis that materially affect the contents of the License Renewal Application, including the UFSAR supplements.	1.4	12 months after LRA submittal date and at least 3 months before completion of NRC review Annual Update submitted by letters dated 4/13/09 and 8/12/10
2	Following the issuance of the renewed operating license, the summary descriptions of aging management programs and TLAAs provided in Appendix A, and the final list of License Renewal commitments, will be incorporated into the PINGP UFSAR as part of a periodic UFSAR update in accordance with 10 CFR 50.71(e). Other changes to specific sections of the PINGP UFSAR necessary to reflect a renewed operating license will also be addressed at that time.	A1.0	First UFSAR update in accordance with 10 CFR 50.71(e) following issuance of renewed operating licenses
3	An Aboveground Steel Tanks Program will be implemented. Program features will be as described in LRA Section B2.1.2.	B2.1.2	U1 - 8/9/2013 U2 - 10/29/2014
4	Procedures for the conduct of inspections in the External Surfaces Monitoring Program, Structures Monitoring Program, Buried Piping and Tanks Inspection Program, and the RG 1.127 Inspection of Water-Control Structures Associated with Nuclear Power Plants Program will be enhanced to include guidance for visual inspections of installed bolting.	B2.1.6	U1 - 8/9/2013 U2 - 10/29/2014
5	A Buried Piping and Tanks Inspection Program will be implemented. Program features will be as described in LRA Section B2.1.8.	B2.1.8	U1 - 8/9/2013 U2 - 10/29/2014
6	The Closed-Cycle Cooling Water System Program will be enhanced to include periodic inspection of accessible surfaces of components serviced by closed-cycle cooling water when the systems or components are opened during scheduled maintenance or surveillance activities. Inspections are performed to identify the presence of aging effects and to confirm the effectiveness of the chemistry controls. Visual inspection of component internals will be used to detect loss of material and heat transfer degradation. Enhanced visual or volumetric examination techniques will be used to detect cracking. [Revised in letter dated 1/20/2009 in response to RAI 3.3.2-13-01]	B2.1.9	U1 - 8/9/2013 U2 - 10/29/2014

A-2

	APPENDIX A: PINGP LICENSE RENEWAL COMMITMENTS		
Commitment Number	Commitment	UFSAR Supplement Section/LRA Section	Enhancement or Implementation Schedule
7	The Compressed Air Monitoring Program will be enhanced as follows: • Station and Instrument Air System air quality will be monitored and maintained in accordance with the instrument air quality guidance provided in ISA S7.0.01-1996. Particulate testing will be revised to use a particle size methodology as specified in ISA S7.0.01. • The program will incorporate on-line dew point monitoring. [Revised in letter dated 2/6/2009 in response to Region III License Renewal Inspection]	B2.1.10	U1 - 8/9/2013 U2 - 10/29/2014
8	An Electrical Cable Connections Not Subject to 10 CFR 50.49 Environmental Qualification Requirements Program will be completed. Program features will be as described in LRA Section B2.1.11.	B2.1.11	U1 - 8/9/2013 U2 - 10/29/2014
9	An Electrical Cables and Connections Not Subject to 10 CFR 50.49 Environmental Qualification Requirements Program will be implemented. Program features will be as described in LRA Section B2.1.12.	B2.1.12	U1 - 8/9/2013 U2 - 10/29/2014
10	An Electrical Cables and Connections Not Subject to 10 CFR 50.49 Environmental Qualification Requirements Used in Instrumentation Circuits Program will be implemented. Program features will be as described in LRA Section B2.1.13.	B2.1.13	U1 - 8/9/2013 U2 - 10/29/2014

Appendix A

APPENDIX A: PINGP LICENSE RENEWAL COMMITMENTS

Commitment Number	Commitment	UFSAR Supplement Section/LRA Section	Enhancement or Implementation Schedule
11	The External Surfaces Monitoring Program will be enhanced as follows: • The scope of the program will be expanded as necessary to include all metallic and non-metallic components within the scope of License Renewal that require aging management in accordance with this program. • The program will ensure that surfaces that are inaccessible or not readily visible during plant operations will be inspected during refueling outages. • The program will ensure that surfaces that are inaccessible or not readily visible during both plant operations and refueling outages will be inspected at intervals that provide reasonable assurance that aging effects are managed such that the applicable components will perform their intended function during the period of extended operation. • The program will apply physical manipulation techniques, in addition to visual inspection, to detect aging effects in elastomers and plastics. • The program will include acceptance criteria (e.g., threshold values for identified aging effects) to ensure that the need for corrective actions will be identified before a loss of intended functions. • The program will ensure that program documentation such as walkdown records, inspection results, and other records of monitoring and trending activities are auditable and retrievable. [Revised in letter dated 2/6/2009 in response to RAI B2.1.14-1 Follow up question]	B2.1.14	U1 - 8/9/2013 U2 - 10/29/2014
12	The Fire Protection Program will be enhanced to require periodic visual inspection of the fire barrier walls, ceilings, and floors to be performed during walkdowns at least once every refueling cycle. [Revised in letter dated 12/5/2008 in response to RAI B2.1.15-3]	B2.1.15	U1 - 8/9/2013 U2 - 10/29/2014

APPENDIX A: PINGP LICENSE RENEWAL COMMITMENTS

Commitment Number	Commitment	UFSAR Supplement Section/LRA Section	Enhancement or Implementation Schedule
13	The Fire Water System Program will be enhanced as follows: • The program will be expanded to include eight additional yard fire hydrants in the scope of the annual visual inspection and flushing activities. • The program will require that sprinkler heads that have been in place for 50 years will be replaced or a representative sample of sprinkler heads will be tested using the guidance of NFPA 25, "Inspection, Testing and Maintenance of Water-Based Fire Protection Systems" (2002 Edition, Section 5.3.1.1.1). Sample testing, if performed, will continue at a 10-year interval following the initial testing.	B2.1.16	U1 - 8/9/2013 U2 - 10/29/2014
14	The Flux Thimble Tube Inspection Program will be enhanced as follows: • The program will require that the interval between inspections be established such that no flux thimble tube is predicted to incur wear that exceeds the established acceptance criteria before the next inspection. • The program will require that re-baselining of the examination frequency be justified using plant-specific wear rate data unless prior plant-specific NRC acceptance for the re-baselining was received. If design changes are made to use more wear-resistant thimble tube materials, sufficient inspections will be conducted at an adequate inspection frequency for the new materials. • The program will require that flux thimble tubes that cannot be inspected must be removed from service.	B2.1.18	U1 - 8/9/2013 U2 - 10/29/2014
15	The Fuel Oil Chemistry Program will be enhanced as follows: • Particulate contamination testing of fuel oil in the 11 fuel oil storage tanks in-scope of License Renewal will be performed, in accordance with ASTM D 6217, on an annual basis. • One-time ultrasonic thickness measurements will be performed at selected tank bottom and piping locations prior to the period of extended operation.	B2.1.19	U1 - 8/9/2013 U2 - 10/29/2014
16	A Fuse Holders Program will be implemented. Program features will be as described in LRA Section B2.1.20.	B2.1.20	U1 - 8/9/2013 U2 - 10/29/2014

APPENDIX A: PINGP LICENSE RENEWAL COMMITMENTS

Commitment Number	Commitment	UFSAR Supplement Section/LRA Section	Enhancement or Implementation Schedule
17	An Inaccessible Medium Voltage Cables Not Subject to 10 CFR 50.49 Environmental Qualification Requirements Program will be implemented. Program features will be as described in LRA Section B2.1.21	B2.1.21	U1 - 8/9/2013 U2 - 10/29/2014
18	An Inspection of Internal Surfaces in Miscellaneous Piping and Ducting Components Program will be implemented. Program features will be as described in LRA section B2.1.22. Inspections for stress corrosion cracking will be performed by visual examination with a magnified resolution as described in 10 CFR 50.55a(b)(2)(xxi)(A) or with ultrasonic methods. [Revised in letter dated 2/6/2009 in response to RAI B2.1.22-1 Follow Up question]	B2.1.22	U1 - 8/9/2013 U2 - 10/29/2014
19	The Inspection of Overhead Heavy Load and Light Load (Related to Refueling) Handling Systems Program will be enhanced as follows: • Program implementing procedures will be revised to ensure the components and structures subject to inspection are clearly identified. • Program inspection procedures will be enhanced to include the parameters corrosion and wear where omitted.	B2.1.23	U1 - 8/9/2013 U2 - 10/29/2014
20	A Metal-Enclosed Bus Program will be implemented. Program features will be as described in LRA Section B2.1.26.	B2.1.26	U1 - 8/9/2013 U2 - 10/29/2014
21	Number Not Used [Deleted by Applicant in a letter Dated 3/27/2009]		
22	Number Not Used [Deleted by Applicant in a letter Dated 4/13/2009]		
23	A One-Time Inspection Program will be completed. Program features will be as described in LRA Section B2.1.29.	B2.1.29	U1 - 8/9/2013 U2 - 10/29/2014

APPENDIX A: PINGP LICENSE RENEWAL COMMITMENTS

Commitment Number	Commitment	UFSAR Supplement Section/LRA Section	Enhancement or Implementation Schedule
24	A. A One-Time Inspection of ASME Code Class 1 Small-Bore Piping Program will be completed prior to the period of extended operation except as noted in Part B of this commitment. Program features will be as described in LRA Section B2.1.30. The following examinations of ASME Code Class 1 small-bore piping socket welds will be performed prior to the period of extended operation: • Volumetric examinations of two socket welds on Unit 1 and three socket welds on Unit 2, or • Destructive examination of two socket welds per Unit. B. Socket weld examinations required by the One-Time Inspection of ASME Code Class 1 Small-Bore Piping Program, not performed prior to the period of extended operation, will be performed within three years of each Unit entering the period of extended operation. [Revised in letter dated 12/17/10 in response to RAI B.2.1.30]	B2.1.30	U1 - 8/9/2013 U2 - 10/29/2014 U1 - 8/9/2016 U2 - 10/29/2017
25	A. A PWR Vessel Internals Program will be implemented. Program features will be as described in LRA Section B2.1.32. B. An inspection plan for reactor internals will be submitted for NRC review and approval at least 24 months prior to the period of extended operation. In addition, the submittal will include any necessary revisions to the PINGP PWR Vessel Internals Program, as well as any related changes to the PINGP scoping, screening and aging management review results for reactor internals, to conform to the NRC-approved Inspection and Evaluation Guidelines. [Revised in letter dated 5/12/2009] [Revised in letter dated 6/24/09 in response to Follow-up RAI B2.1.38]	B2.1.32	U1 – 8/9/2013 U2 – 10/29/2014 U1 – 8/9/2011 U2 – 10/29/2012
26	The Reactor Head Closure Studs Program will be enhanced to incorporate controls that ensure that any future procurement of reactor head closure studs will be in accordance with the material and inspection guidance provided in NRC Regulatory Guide 1.65.	B2.1.33	U1 - 8/9/2013 U2 - 10/29/2014

Appendix A

	APPENDIX A: PINGP LICENSE RENEWAL COMMITMENTS		
Commitment Number	Commitment	UFSAR Supplement Section/LRA Section	Enhancement or Implementation Schedule
27	The Reactor Vessel Surveillance Program will be enhanced as follows: • A requirement will be added to ensure that all withdrawn and tested surveillance capsules, not discarded as of August 31, 2000, are placed in storage for possible future reconstitution and use. • A requirement will be added to ensure that in the event spare capsules are withdrawn, the untested capsules are placed in storage and maintained for future insertion.	B2.1.34	U1 - 8/9/2013 U2 - 10/29/2014
28	The RG 1.127, Inspection of Water-Control Structures Associated with Nuclear Power Plants Program will be enhanced as follows: • The program will include inspections of concrete and steel components that are below the water line at the Screenhouse and Intake Canal. The scope will also require inspections of the Approach Canal, Intake Canal, Emergency Cooling Water Intake, and Screenhouse immediately following extreme environmental conditions or natural phenomena including an earthquake, flood, tornado, severe thunderstorm, or high winds. • The program parameters to be inspected will include an inspection of water-control concrete components that are below the water line for cavitation and erosion degradation. • The program will visually inspect for damage such as cracking, settlement, movement, broken bolted and welded connections, buckling, and other degraded conditions following extreme environmental conditions or natural phenomena.	B2.1.35	U1 - 8/9/2013 U2 - 10/29/2014
29	A Selective Leaching of Materials Program will be completed. Program features will be as described in LRA B2.1.36.	B2.1.36	U1 - 8/9/2013 U2 - 10/29/2014

A-8

APPENDIX A: PINGP LICENSE RENEWAL COMMITMENTS

Commitment Number	Commitment	UFSAR Supplement Section/LRA Section	Enhancement or Implementation Schedule
30	The Structures Monitoring Program will be enhanced as follows: • The following structures, components, and component supports will be added to the scope of the inspections: - Approach Canal - Fuel Oil Transfer House - Old Administration Building and Administration Building Addition - Component supports for cable tray, conduit, cable, tubing tray, tubing, non-ASME vessels, exchangers, pumps, valves, piping, mirror insulation, non-ASME valves, cabinets, panels, racks, equipment enclosures, junction boxes, bus ducts, breakers, transformers, instruments, diesel equipment, housings for HVAC fans, louvers, and dampers, HVAC ducts, vibration isolation elements for diesel equipment, and miscellaneous electrical and mechanical equipment items - Miscellaneous electrical equipment and instrumentation enclosures including cable tray, conduit, wireway, tube tray, cabinets, panels, racks, equipment enclosures, junction boxes, breaker housings, transformer housings, lighting fixtures, and metal bus enclosure assemblies - Miscellaneous mechanical equipment enclosures including housings for HVAC fans, louvers, and dampers - SBO Yard Structures and components including SBO cable vault and bus duct enclosures - Fire Protection System hydrant houses - Caulking, sealant and elastomer materials - Nonsafety-related masonry walls that support equipment relied upon to perform a function that demonstrates compliance with a regulated event(s). • The program will be enhanced to include additional inspection parameters. • The program will require an inspection frequency of once every five (5) years for structures and structural components within the scope of the program. The frequency of inspections can be adjusted, if necessary, to allow for early detection and timely correction of negative trends. • The program will require periodic sampling of groundwater and river water chemistries to ensure they remain non-aggressive.	B2.1.38	U1 - 8/9/2013 U2 - 10/29/2014

APPENDIX A: PINGP LICENSE RENEWAL COMMITMENTS

Commitment Number	Commitment	UFSAR Supplement Section/LRA Section	Enhancement or Implementation Schedule
31	A Thermal Aging Embrittlement of Cast Austenitic Stainless Steel (CASS) Program will be implemented. Program features will be as described in LRA Section B2.1.39.	B2.1.39	U1 - 8/9/2013 U2 - 10/29/2014
32	The Water Chemistry Program will be enhanced as follows: • The program will require increased sampling to be performed as needed to confirm the effectiveness of corrective actions taken to address an abnormal chemistry condition. • The program will require Reactor Coolant System dissolved oxygen Action Level limits to be consistent with the limits established in the EPRI PWR Primary Water Chemistry Guidelines. [Revised in letter dated 12/5/2008 in response to RAI B2.1.40-3]	B2.1.40	U1 - 8/9/2013 U2 - 10/29/2014
33	The Metal Fatigue of Reactor Coolant Pressure Boundary Program will be enhanced as follows: • The program will monitor the six component locations identified in NUREG/CR-6260 for older vintage Westinghouse plants, either by tracking the cumulative number of imposed stress cycles using cycle counting, or by tracking the cumulative fatigue usage, including the effects of coolant environment. The following locations will be monitored: - Reactor Vessel Inlet and Outlet Nozzles - Reactor Pressure Vessel Shell to Lower Head - RCS Hot Leg Surge Line Nozzle - RCS Cold Leg Charging Nozzle - RCS Cold Leg Safety Injection Accumulator Nozzle - RHR-to-Accumulator Piping Tee • Program acceptance criteria will be clarified to require corrective action to be taken before a cumulative fatigue usage factor exceeds 1.0 or a design basis transient cycle limit is exceeded [Revised in letter dated 1/9/2009 in response to RAI 4.3.1.1-1]	B3.2	U1 - 8/9/2013 U2 - 10/29/2014
34	Reactor internals baffle bolt fatigue transient limits of 1835 cycles of plant loading at 5% per minute and 1835 cycles of plant unloading at 5% per minute will be incorporated into the Metal Fatigue of Reactor Coolant Pressure Boundary Program and UFSAR Table 4.1-8.	B3.2	U1 - 8/9/2013 U2 - 10/29/2014

APPENDIX A: PINGP LICENSE RENEWAL COMMITMENTS

Commitment Number	Commitment	UFSAR Supplement Section/LRA Section	Enhancement or Implementation Schedule
35	NSPM will perform an ASME Section III fatigue evaluation of the lower head of the pressurizer to account for effects of insurge/outsurge transients. The evaluation will determine the cumulative fatigue usage of limiting pressurizer component(s) through the period of extended operation. The analyses will account for periods of both "Water Solid" and "Standard Steam Bubble" operating strategies. Analysis results will be incorporated, as applicable, into the Metal Fatigue of Reactor Coolant Pressure Boundary Program. [Revised in letter dated 1/9/2009 in response to RAI 4.3.1.1-1]	4.3.1.3	U1 - 8/9/2013 U2 - 10/29/2014
36	NSPM will complete fatigue calculations for the pressurizer surge line hot leg nozzle and the charging nozzle using the methodology of the ASME Code (Subsection NB) and will report the revised CUFs and CUFs adjusted for environmental effects at these locations as an amendment to the PINGP LRA. Conforming changes to LRA Section 4.3.3, "PINGP EAF Results," will also be included in that amendment to reflect analysis results and remove references to stress-based fatigue monitoring. [Added in letter dated 1/9/2009 in response to RAI 4.3.1.1-1]	4.3.3	4/30/2009 Letter dated 4/28/2009 from the applicant to NRC completes this commitment, see ML091190418
37	NSPM will revise procedures for excavation and trenching controls and archaeological, cultural and historic resource protection to identify sensitive areas and provide guidance for ground-disturbing activities. The procedures will be revised to include drawings and illustrations to assist users in identifying culturally sensitive areas, and pictures of artifacts that are prevalent in the area of the Plant site. The revised procedures will also require training of the Site Environmental Coordinator and other personnel responsible for proper execution of excavation or other ground-disturbing activities. [Added in ER revision submitted in letter dated 3/4/2009]	ER 4.16.1	8/9/2013
38	NSPM will conduct a Phase I Reconnaissance Field Survey of the disturbed areas within the Plant's boundaries. In addition, NSPM will conduct Phase I field surveys of areas of known archaeological sites to precisely determine their boundaries. NSPM will use the results of these surveys to designate areas for archaeological protection. [Added in ER revision submitted in letter dated 3/4/2009]	ER 4.16.2	8/9/2013

Appendix A

APPENDIX A: PINGP LICENSE RENEWAL COMMITMENTS

Commitment Number	Commitment	UFSAR Supplement Section/LRA Section	Enhancement or Implementation Schedule
39	NSPM will prepare, maintain and implement a Cultural Resources Management Plan (CRMP) to protect significant historical, archaeological, and cultural resources that may currently exist on the Plant site. In connection with the preparation of the CRMP, NSPM will conduct botanical surveys to identify culturally and medicinally important species on the Plant site, and incorporate provisions to protect such plants into the CRMP. [Added in ER revision submitted in letter dated 3/4/2009]	ER 4.16.2	8/9/2013
40	NSPM will consult with a qualified archaeologist prior to conducting any ground-disturbing activity in any area designated as undisturbed and in any disturbed area that is described as potentially containing archaeological resources (as determined by the Phase I Reconnaissance Field Survey discussed in Commitment Number 38). [Added in ER revision submitted in letter dated 3/4/2009]	ER 4.16.2	8/9/2013
41	During the first refueling outage following refueling cavity leak repairs in each Unit (scheduled for refueling outages 1R26 and 2R26), concrete will be removed from the Sump C pit to expose an area of the containment vessel bottom head. Visual examination and ultrasonic thickness measurement will be performed on the portions of the containment vessels exposed by the excavations. An assessment of the condition of exposed concrete and rebar will also be performed. Petrographic examination will be performed on sample pieces of the removed concrete if the removal method provides pieces suitable for examination. Degradation observed in the exposed containment vessel, concrete or rebar, or as a result of petrographic examination of concrete samples, will be entered into the Corrective Action Program, and evaluated for impact on structural integrity and identification of additional actions that may be warranted. [Added in letter dated 4/6/09 in response to Follow Up RAI B2.1.38] [Revised in letter dated 8/7/09 in response to a follow-up question from a conference call on 7/22/09]	B2.1.38	U1 - 8/9/2013 U2 - 10/29/2014

A-12

APPENDIX A: PINGP LICENSE RENEWAL COMMITMENTS

Commitment Number	Commitment	UFSAR Supplement Section/LRA Section	Enhancement or Implementation Schedule
42	During the two consecutive refueling outages following refueling cavity leak repairs in each Unit (scheduled for refueling outages 1R26 and 2R26), visual inspections will be performed of the areas where reactor cavity leakage had been observed previously to confirm that leakage has been resolved. The inspection results will be documented. If refueling cavity leakage is again identified, the issue will be entered into the Corrective Action Program and evaluated for identification of additional actions to mitigate leakage and monitor the condition of the containment vessel and internal structures. [Added in letter dated 4/6/09 in response to Follow Up RAI B2.1.38]	B2.1.38	U1 - 8/9/2013 U2 - 10/29/2014
43	Preventive maintenance requirements will be implemented to require periodic replacement of rubber flexible hoses in the Diesel Generators and Support System and in the 122 Diesel Driven Fire Pump that are exposed to fuel oil or lubricating oil internal environments. [Added in letter dated 4/6/09 in response to RAI 3.3.2-8-1] [Revised in letter dated 6/5/09]	Table 3.3.2-8	U1 - 8/9/2013 U2 - 10/29/2014
44	During the first refueling outage following refueling cavity leak repairs in each Unit (scheduled for refueling outages 1R26 and 2R26), a concrete sample will be obtained from a location known to have been wetted by borated water leakage from the refueling cavity. These concrete samples (one per Unit) will be tested for compression strength and will be subjected to petrographic examination to assess the degradation, if any, resulting from borated water exposure. Degradation identified as a result of the testing and examination of the concrete samples will be entered into the Corrective Action Program, and evaluated for impact on structural integrity and identification of additional actions that may be warranted. [Added in letter dated 8/7/09 in response to a follow-up question from a conference call on 7/22/09.]	B2.1.38	U1 - 8/9/2013 U2 - 10/29/2014
45	If the original PINGP Unit 2 steam generators are not replaced prior to entry into the period of extended operation, NSPM will perform an inspection of each PINGP Unit 2 steam generator, prior to the period of extended operation, to assess the condition of the divider plates and associated welds. The examination technique(s) will be capable of detecting PWSCC in the divider plates and associated welds. [Added in letter dated 11/5/10]	None	U2 - 10/29/2014

Appendix A

APPENDIX A: PINGP LICENSE RENEWAL COMMITMENTS

Commitment Number	Commitment	UFSAR Supplement Section/LRA Section	Enhancement or Implementation Schedule
46	A one-time inspection of a representative number of tube-to-tubesheet welds in each Unit 1 steam generator will be performed to determine if primary water stress corrosion cracking (PWSCC) is present. The tube-to-tubesheet weld inspections will be performed during the first Unit 1 refueling outage after the Unit 1 steam generators have reached 20 years of service. If weld cracking is identified: The condition will be resolved through repair or engineering evaluation to justify continued service, as appropriate, and An aging management program will be established to perform routine tube-to-tubesheet weld inspections for the remaining life of the Unit 1 and Unit 2 replacement steam generators. [Added in letter dated 12/17/10 in response to RAI 3.1.2.2.16]	None	Unit 1 steam generator tube-to-tubesheet weld inspection will be performed during first Unit 1 refueling outage after the Unit 1 steam generators have reached 20 years of service.
47	NSPM will perform a review of the design basis ASME Class 1 fatigue evaluations to determine whether the NUREG/CR-6260 components that have previously been evaluated for the effects of reactor coolant environment on fatigue life are the limiting components for the PINGP design. If a more limiting component(s) is identified, the most limiting component will be evaluated for the effects of the reactor coolant environment on fatigue usage. If the limiting component identified consists of nickel alloy, the methodology used to perform the environmentally-assisted fatigue calculation for nickel alloy will be consistent with NUREG/CR-6909, or otherwise justified. [Added in letter dated 12/17/10 in response to RAI 4.3.3]	4.3.3	8/9/2013

A-14

APPENDIX B

CHRONOLOGY

This appendix contains a chronological listing of the licensing correspondence between the staff of the U.S. Nuclear Regulatory Commission and Northern States Power Company, a Minnesota Corporation (NSPM). This appendix updates the correspondence regarding the staff's review of the Prairie Island Nuclear Generating Plant, Units 1 and 2, license renewal application (under Docket Nos. 50-282 and 50-306) since the issuance of the final safety evaluation report in October 2009.

APPENDIX B: CHRONOLOGY	
Date	Subject
12/10/2009	Letter from Mario V. Bonaca, Advisory Committee on Reactor Safeguards, to Gregory B. Jaczko, Chairman NRC, "Report on the Safety Aspects of the License Renewal Application for the Prairie Island Nuclear Generating Plant, Units 1 and 2" (Agencywide Documents Access and Management System [ADAMS] Accession No. ML093420316).
8/12/2010	Letter from Northern States Power Company to NRC, "Annual Update of the Application for Renewed Operating Licenses and Supplemental Information Regarding Buried Piping and Tanks Inspection Program and Class 1 Small-Bore Piping Program" (ADAMS Accession No. ML102250265).
11/5/2010	Letter from Northern States Power Company to NRC, "Supplemental Information Regarding Inaccessible Medium Voltage Cable Program, Buried Piping and Tanks Inspection Program, Inspection of Steam Generator Divider Plates and Class 1 Small-Bore Piping Program" (ADAMS Accession No. ML103130368).
12/17/2010	Letter from Northern States Power Company to NRC, "License Renewal Application, Response to Requests for Additional Information" (TAC Nos. MD8528 and MD8529) (ADAMS Accession No. ML103510501).

APPENDIX C

PRINCIPAL CONTRIBUTORS

This appendix lists the principal contributors for the development of this supplemental safety evaluation report and their areas of responsibility.

APPENDIX C: PRINCIPAL CONTRIBUTORS	
NAME	RESPONSIBILITY
R. Auluck	Management Oversight
C. Doutt	Electrical Engineer
A. Dias	Management Oversight
A. Erickson	General Engineer
B. Fu	Mechanical Engineer
M. Galloway	Management Oversight
A. Hiser	Management Oversight
B. Holian	Management Oversight
W. Holston	Mechanical Engineer
M. Kichline	Mechanical Engineer
B. Lehman	Civil & Structural Engineer
R. Li	Electrical Engineer
C. Ng	Mechanical Engineer
D. Pelton	Management Oversight
R. Plasse	Project Manager
R. Vaucher	Mechanical Engineer
D. Wrona	Management Oversight
O. Yee	Mechanical Engineer

NRC FORM 335 (12-2010) NRCMD 3.7	U.S. NUCLEAR REGULATORY COMMISSION	1. REPORT NUMBER (Assigned by NRC, Add Vol., Supp., Rev., and Addendum Numbers, if any.)
	BIBLIOGRAPHIC DATA SHEET *(See instructions on the reverse)*	NUREG-1960, Vol. 2

2. TITLE AND SUBTITLE

Safety Evaluation Report Related to the License Renewal of Prairie Island Nuclear Generating Plant Units 1 and 2
Supplement 1

3. DATE REPORT PUBLISHED

MONTH	YEAR
August	2011

4. FIN OR GRANT NUMBER

5. AUTHOR(S)

See SER Appendix C

6. TYPE OF REPORT

Final

7. PERIOD COVERED *(Inclusive Dates)*

8. PERFORMING ORGANIZATION - NAME AND ADDRESS *(If NRC, provide Division, Office or Region, U.S. Nuclear Regulatory Commission, and mailing address; if contractor, provide name and mailing address.)*

Division of License Renewal
Office of Nuclear Reactor Regulation
U.S. Nuclear Regulatory Commission
Washington, DC 20555-0001

9. SPONSORING ORGANIZATION - NAME AND ADDRESS *(If NRC, type "Same as above"; if contractor, provide NRC Division, Office or Region, U.S. Nuclear Regulatory Commission, and mailing address.)*

Same as above

10. SUPPLEMENTARY NOTES

Richard Plasse, NRC Project Manager

11. ABSTRACT *(200 words or less)*

This document is a supplemental safety evaluation report (SSER) for the license renewal application (LRA) for Prairie Island Nuclear Generating Plant (PINGP), Units 1 and 2 as submitted by Nuclear Management Company (NMC), LLC. By letter dated April 11, 2008, NMC submitted its LRA to the United States (US) Nuclear Regulatory Commission (NRC) for renewal of the PINGP operating licenses for an additional 20 years. The NRC staff (the staff) issued a Safety Evaluation Report (SER) Related to the License Renewal of Prairie Island Nuclear Generating Plant, Units 1 and 2, dated October 16, 2009, (ADAMS accession number ML09280209) which summarizes the results of the review of the LRA for compliance with the requirements of Title 10 Part 54, of the Code of Federal Regulations, (10 CFR Part 54), "Requirements for Renewal of Operating Licenses for Nuclear Power Plants."

This SSER documents the staff's review of supplemental information provided by the applicant since the issuance of the SER. This information includes annual updates required by 10 CFR 54.21(b), and updated information and commitments in response to the recent industry operating experience.

12. KEY WORDS/DESCRIPTORS *(List words or phrases that will assist researchers in locating the report.)*

10 CFR Part 54
License Renewal
Prairie Island Nuclear Generating Plant
Time-limited aging analysis
Aging Management
Scoping and Screening

13. AVAILABILITY STATEMENT

unlimited

14. SECURITY CLASSIFICATION

(This Page)

unclassified

(This Report)

unclassified

15. NUMBER OF PAGES

16. PRICE

NUREG-1960
Supplement 1

Safety Evaluation Report Related to the License Renewal of
Prairie Island Nuclear Generating Plant Units 1 and 2

August 2011